Grades K-2
ID43012

Balancing Numbers

By Jill Osofsky

Contents

Art Director: Sara Mordecai
Cover Concept and Page Design: Sara Mordecai
Cover Illustration: Kristin Mallory
Technical Art: Mark Keli'ihanapule
Text Illustrations: Jim Connolly
Editor: Nancy Tune
Project Coordinator: Judy Crum

ISBN: 1-56451-318-1
Funtastic Frogs™: Balancing Numbers, Grades K–2

A Division of Instructional Fair Group, Inc.
A Tribune Education Company
3195 Wilson Drive NW, Grand Rapids, MI 49544 • USA
Duke Street, Wisbech, Cambs, PE13 2AE • UK

This book is one in a series of books designed to develop children's mathematical thinking. Each book supports the use of Funtastic Frog™ Counters in activities that teach key mathematical concepts.

Funtastic Frog™ Counters are available from Ideal School Supply in six different colors and in three different sizes (3g, 6g, and 12g). A unique lacing feature allows children to use the frogs in a wide range of counting and patterning activities.

In *Funtastic Frogs™: Balancing Numbers, Grades K-2*, children use the frog counters in activities to:

- represent numerical situations using concrete objects
- work with variables, numbers, and equations in an informal way
- work with the order property of addition in an informal way
- develop number sense
- represent the same number in different ways
- understand the meaning of addition
- find sums through 15
- show equal sums using different numbers
- compare numbers to find which is *more than*, *less than*, and *the same as*
- use appropriate representations of mathematical situations
- solve problems using logical reasoning

The activities in this book are designed to teach different ways of representing equal values. All the activities support current mathematics standards.

As children engage in these activities, they will model equal values using the frog counters and a simple balance. They will develop the understanding that equal values can be shown in different ways and that the order of numbers does not affect the sum. The activities will help children think flexibly about numbers and will help build a foundation for understanding variables and equations when they encounter formal algebra instruction in later years.

Contents

This book contains 22 activities divided into four sections: finding equal values, comparing number relationships to find which is *more than*, comparing number relationships to find which is *less than*, and comparing number relationships to find which is *more than*, *less than*, and *the same as*. The activities in each section are similar, giving the children the opportunity to practice and use each skill.

Math Skills and Understandings	**Related Activities**
Represent numbers using concrete objects	Activities 1–22
Work with variables, numbers, and equations in an informal way	Activities 1–22
Represent numbers in different ways	Activities 1–22
Understand the meaning of addition	Activities 1–22
Find sums through 15	Activities 1–22
Find equal sums using different numbers	Activities 1–22
Compare values to find which is *more than*, *less than*, and *the same as*	Activities 10–22
Use appropriate representations of mathematical situations	Activities 1–22
Use logic and reasoning to solve problems	Activities 1–22

Suggestions for Classroom Use

The activities in this book are sequenced by level of difficulty within each section and from section to section. Modify a section if you find it is too challenging for your children, or not challenging enough.

The activities can be introduced to the whole class using an overhead projector, the chalkboard, or sitting in a circle on the floor. Once children understand the directions, they can work in pairs, small groups, or individually at a learning center.

Encourage children to share their thinking with the whole class. When working together, they often discover multiple ways to solve a problem. Talking about their thinking and discoveries helps them clarify their thoughts and allows others to hear how they might solve the same problem a different way. Talking about how they solve problems helps children make mathematical connections and deepens their understanding. Class discussions encourage the type of thinking required for reasoning and problem solving.

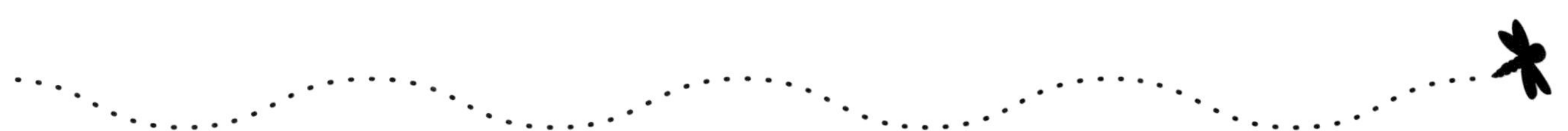

Materials

You may wish to make an overhead transparency of the first activity page in each section as an introduction to the activities that follow.

For each child, pair of children, or group you will need:

- pencils
- crayons or colored markers to match the colors of the frogs
- a tub of frogs in three sizes and six colors
- an activity sheet for each child or pair of children
- a balance for the whole class or each group

Introducing the Activities

Before beginning instruction, it is important to allow time for children to freely explore the frog counters. This will allow them to satisfy their curiosity about the frogs before they use them in structured activities. Encourage the children to compare the frogs and tell how they are alike and how they are different. Ask them to describe the different sizes and weights. The frog sizes are referred to on the activity pages as *small*, *medium*, and *large*. Be sure all children become familiar with those terms before beginning instruction.

Allow groups to explore using the frogs with the balance. You can use this exploration period to introduce vocabulary that will be used throughout the book, such as *balance*, *left*, *right*, and *value*.

When you introduce each section, lead the children through the first activity in that section. Discuss the directions and how to record their work. After you have completed placing the frogs, point to the frogs and have the students count the values with you aloud.

After the children complete the activities, have them talk about how they showed the numbers. Help them extend their thinking by asking such questions as: *Which values were easiest to show? Which were hardest? How can you be sure that the frogs you used on the balance were correct?* After children have worked through the first section, ask more probing questions such as: *What do you notice about the number sentences? Did you find any patterns? What number can be made using the most number combinations?* Encourage children who finish early to look for number relationships for numbers through 20 and beyond.

Mathematical Content by Section

Activities 1–10: Find equal values. To introduce Activities 1–5, tell the children that each small frog has a value of 1. Show five small frogs and ask children to find the total value. Repeat the activity several times using different numbers of small frogs. Next tell the children that each medium frog has a value of 2. Show five medium frogs and ask children to find the total value. Repeat this several times, using different numbers of medium frogs.

Show a combination of small and medium frogs and ask children to find the total value. Ask different children to come up and explain how to find the total. If necessary, show the children how to count-on by one or two for each frog to find the total value.

Use the frog counters and the balance to demonstrate a number sentence that shows the same value on each side. Put four small frogs on the left side of the balance. Ask: *How many small frogs do I need on the other side to balance? Can you show me the same value using the medium frogs?* As volunteers place medium frogs on the right side to balance, have them explain their reasoning. Point out that there is more than one way to balance the value on the left. Show the children how to write the balanced number sentences that have been demonstrated, using an equal sign to represent the middle of the balance.

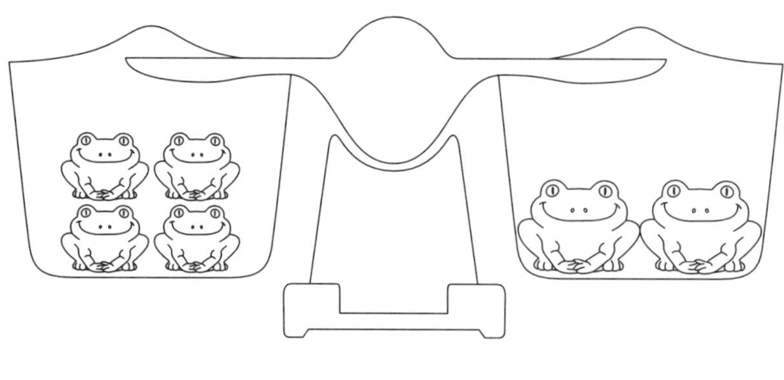

$$1 + 1 + 1 + 1 = 2 + 2$$

To introduce Activity 6, show a large frog and explain that it has a value of 4. Ask children to find the total value of one small, one medium, and one large frog (7). Then make different combinations of frogs for them to solve. Demonstrate using the balance to solve problems with the three sizes of frogs before assigning Activities 6–10.

Activities 11–15: Comparing number relationships to find which is *more than*. Children apply what they have learned about writing balanced number sentences to find which side of the balance has a higher value, or *more than*. To introduce the activities, tell the children that each large frog has a value of 4. Place three large frogs on the left side of the balance and one large frog on the right. Ask the children to tell, by looking at the balance, which side has more. Have them explain their reasoning, making sure they note the relationship between the position of the balance and the value of the numbers—with the larger number value pushing down its side of the balance. Repeat, using different frogs and number values. Model showing the relationship, using the number sentence on page 21.

Activities 16–20: Comparing number relationships to find which is *less than*. Children apply what they have learned about comparing numbers to find which side of the balance has *less*. To introduce the activities, tell the children that each small frog has a value of 1. Place two small frogs on the left side of the balance and one large frog on the right. Ask the children to tell, by looking at the balance, which side has less. Have children explain their reasoning, making sure they note the relationship between the position of the balance (with the higher number value "weighing" more) and the value of the numbers. Repeat, using different frogs and number values. Model how children can show the relationship, using the number sentence on page 26.

Activities 21–22: Comparing number relationships to find which is *more than, less than*, and *the same as*. Introduce these activities by asking children to imagine that the frogs on each side are placed on the balance. The children will then apply what they have learned to determine what kind of number relationship is shown on the balance.

Sample Solutions

These are sample solutions. Additional solutions are possible for some activities.

Activity

1A 4 or 1+1+1+1
1B 1+1+1+1=2+2
EXPLORE MORE 1+1+1+1=1+1+2

2A 2+2+2
2B 2+2+2=1+1+1+1+1+1
EXPLORE MORE 2+2+2=1+1+1+1+2

3A 2+2+1+1+1
3B 2+2+1+1+1=2+2+2+1
EXPLORE MORE 2+2+1+1+1=2+1+1+1+1+1

4A 2+2+2+2
4B 2+2+2+2=2+2+2+1+1
EXPLORE MORE 2+2+2+2=2+2+1+1+1+1

5A 2+2+2+2+1
5B 2+2+2+2+1=2+2+2+1+1+1
EXPLORE MORE 2+2+2+2+1=2+2+1+1+1+1+1

6A 2+2+2+2+2
6B 2+2+2+2+2=2+2+2+1+1+1+1
EXPLORE MORE 2+2+2+2+2=2+2+1+1+1+1+1+1

7A 6
7B 6=3+3
EXPLORE MORE 6=6

8A 3+3+3
8B 3+3+3=3+6
EXPLORE MORE 3+3+3=6+3

9A 12
9B 12=6+6
EXPLORE MORE 12=3+3+6

10A 6+6+3
10B 6+6+3=3+12
EXPLORE MORE 6+6+3=6+3+6

11A 8
11B 8 is more than 2+2

Activity

12A 2+8
12B 2+8 is more than 2+2+2
EXPLORE MORE 2+8 is more than 8

13A 4+8
13B 4+8 is more than 2+8
EXPLORE MORE 4+8 is more than 4+4

14A 2+4+4
14B 2+4+4 is more than 2+4
EXPLORE MORE 2+4+4 is more than 2+2+4

15A 2+2+2+2+2+2
15B 2+2+2+2+2+2 is more than 8+2
EXPLORE MORE 2+2+2+2+2+2 is more than 2+2+4

16A 4
16B 8
16C 4 is less than 8
EXPLORE MORE 4 is less than 2+4

17A 8
17B 4
17C 4 is less than 8
EXPLORE MORE 2+4 is less than 8

18A 2+8
18B 2+4
18C 2+4 is less than 2+8
EXPLORE MORE 2+2+4 is less than 10

19A 2+4+8
19B 4+8
19C 4+8 is less than 2+4+8
EXPLORE MORE 2+8 is less than 2+4+8

20A 2+2+2+2+2
20B 4
20C 4 is less than 2+2+2+2+2
EXPLORE MORE 8 is less than 2+2+2+2+2

21A 2+2
21B 4
21C 2+2 is the same as 4

22A 8+2
22B 4+4
22C 2+8 is more than 4+4

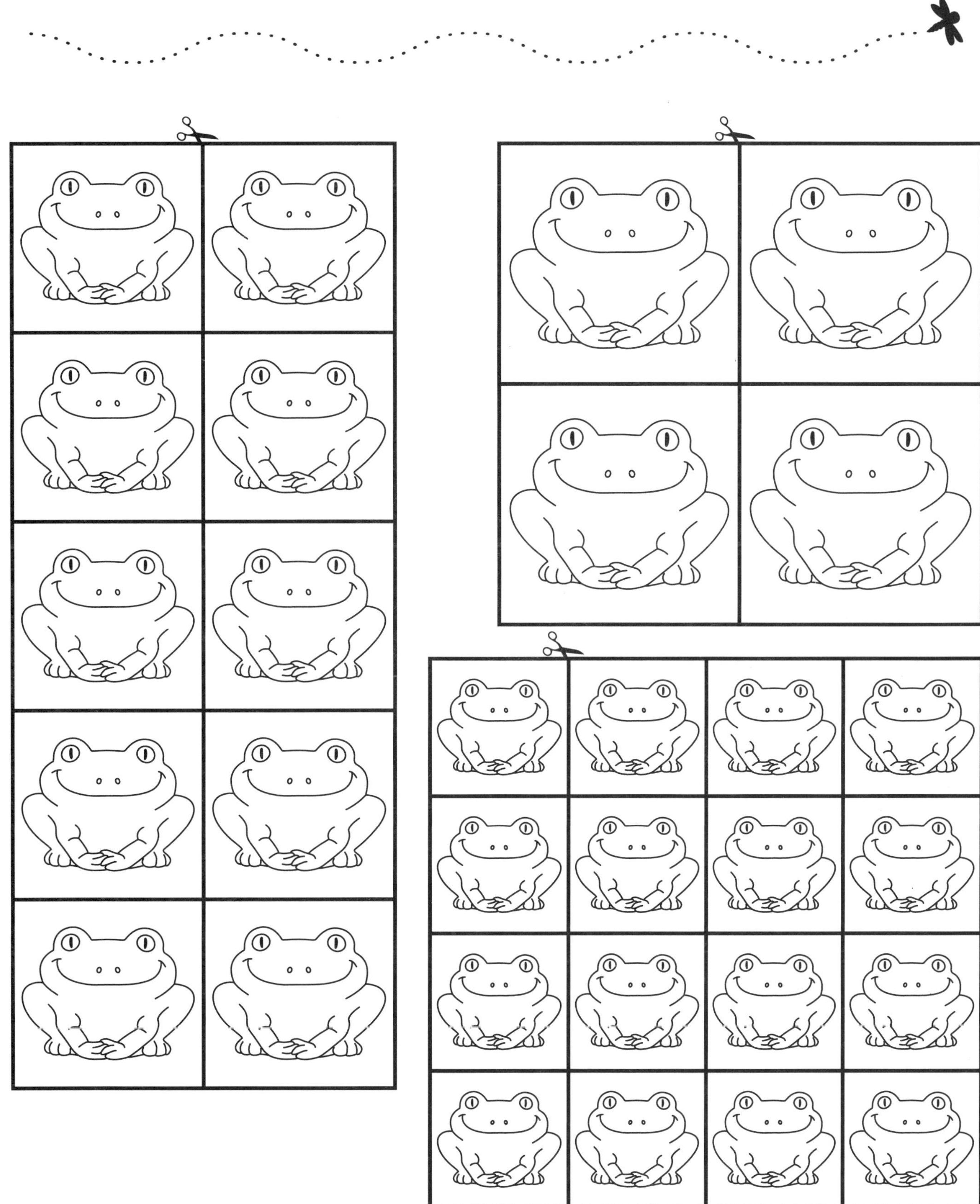

This page may be used to create your own activity sheets.

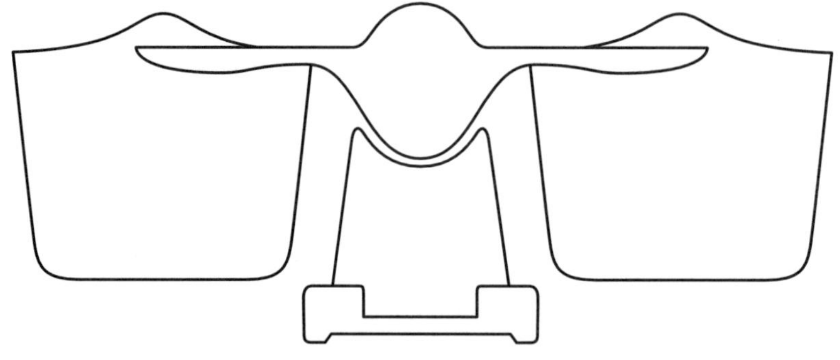

__________ **is the same as** __________

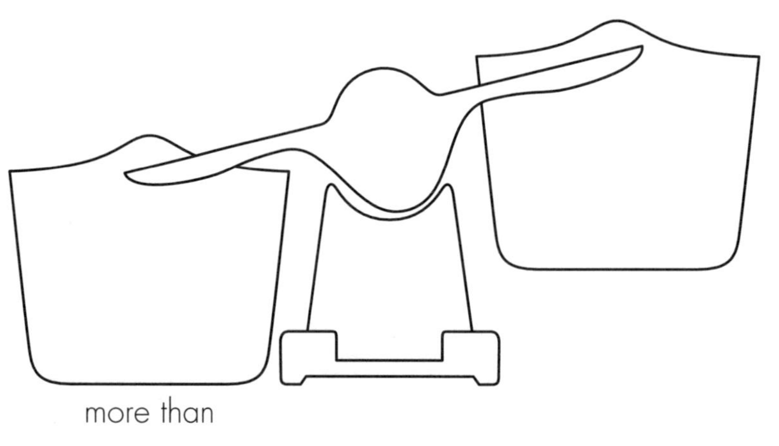

__________ **is more than** __________

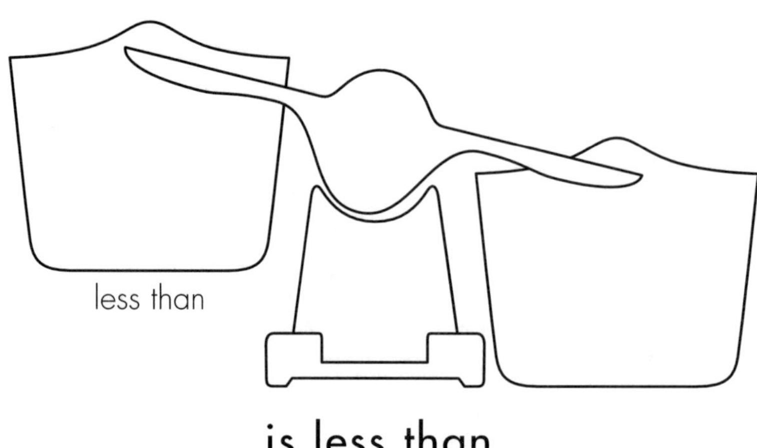

__________ **is less than** __________

Find equal values

Name ____________________

Use small and medium frogs.
Record you work.

Each small frog has a value of 1.

Each medium frog has a value of 2.

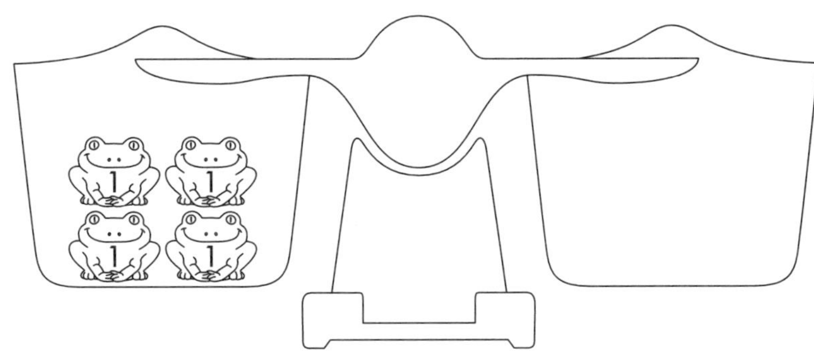

A. What is the value of the frogs on the left side of the balance? __________

Put medium frogs on the right side to balance.
The right side must have the same value as the left side.

B. Write a balanced number sentence.

1 + 1 + 1 + 1 = __________

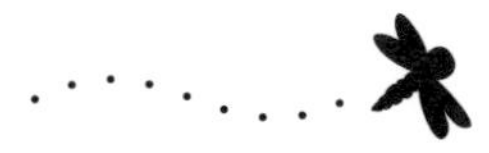

EXPLORE MORE

Find another way to balance the number sentence.

Find equal values

Name ________________________

Use small and medium frogs.
Record your work.

Each small frog has a value of 1.

Each medium frog has a value of 2.

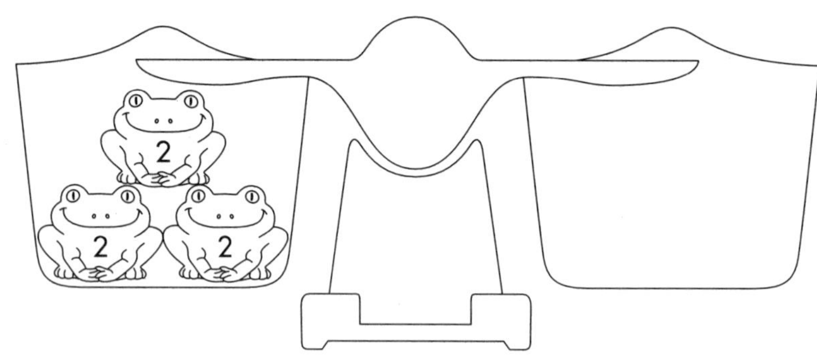

A. What is the value of the frogs on the left side of the balance? __________

Put small frogs on the right side to balance.
The right side must have the same value as the left side.

B. Write a balanced number sentence.

$2 + 2 + 2 =$ __________

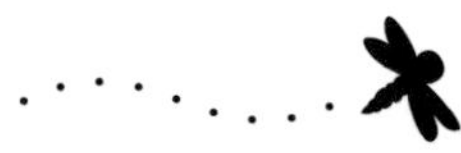

EXPLORE MORE

Find another way to balance the number sentence.

Find equal values

Name ____________________

Use small and medium frogs.
Record your work.

Each small frog has a value of 1.

Each medium frog has a value of 2.

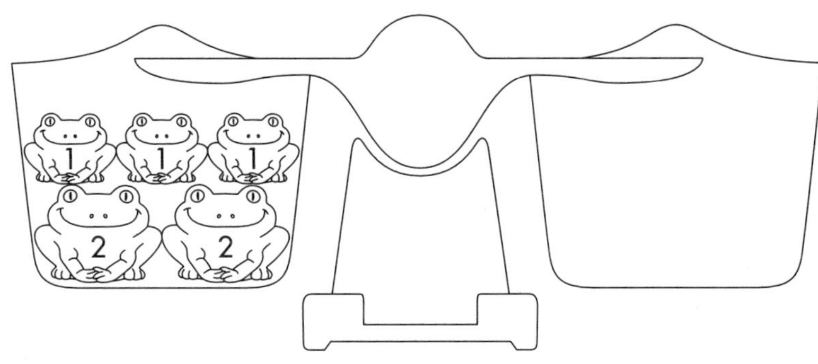

A. What is the value of the frogs on the left side of the balance? __________

Put frogs on the right side to balance.
The right side must have the same value as the left side.

B. Write a balanced number sentence.

2 + 2 + 1 + 1 + 1 = __________

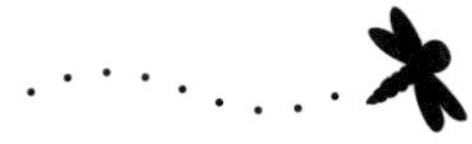

EXPLORE MORE

Find another way to balance the number sentence.

Find equal values

Name ______________________

Use small and medium frogs.
Record your work.

Each small frog has a value of 1.

Each medium frog has a value of 2.

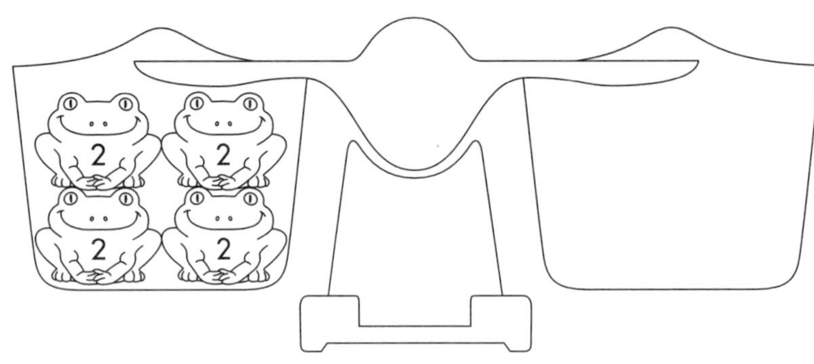

A. What is the value of the frogs on the left side of the balance? __________

Put frogs on the right side to balance.
The right side must have the same value as the left side.

B. Write a balanced number sentence.

2 + 2 + 2 + 2 = __________

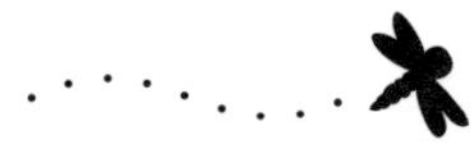

EXPLORE MORE

Find another way to balance the number sentence.

Find equal values

Name ______________________

Use small and medium frogs.
Record your work.

Each small frog has a value of 1.

Each medium frog has a value of 2.

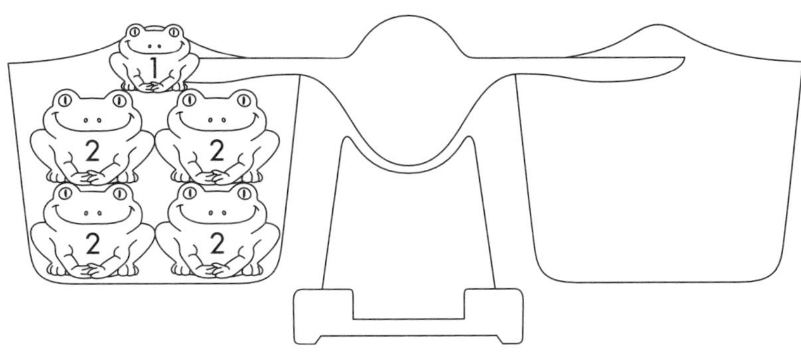

A. What is the value of the frogs on the left side of the balance? __________

Put frogs on the right side to balance.
The right side must have the same value as the left side.

B. Write a balanced number sentence.

2 + 2 + 2 + 2 + 1 = __________

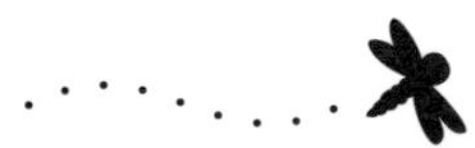

EXPLORE MORE

Find another way to balance the number sentence.

Find equal values

Name ____________________

Use small, medium, and large frogs.
Record your work.

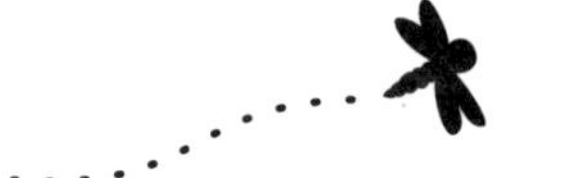

Each small frog has a value of 1.

Each medium frog has a value of 2.

Each large frog has a value of 4.

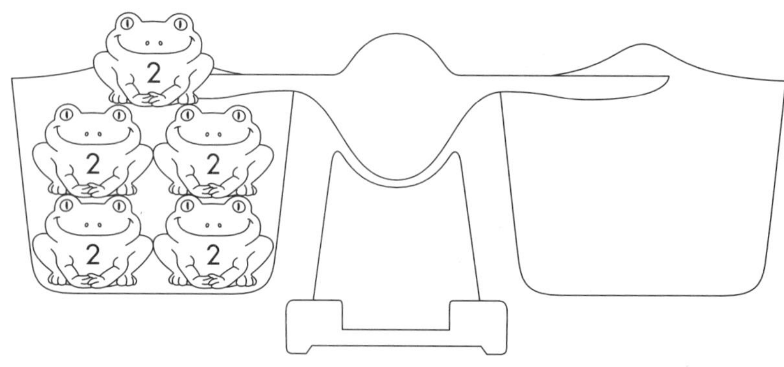

A. What is the value of the frogs on the left side of the balance? __________

Put frogs on the right side to balance.
The right side must have the same value as the left side.

B. Write a balanced number sentence.

2 + 2 + 2 + 2 + 2 = __________

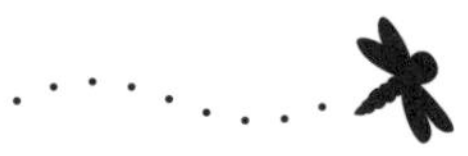

EXPLORE MORE

Find another way to balance the number sentence.

Find equal values

Name ____________________

Use small and medium frogs.
Record your work.

Each small frog has a value of 3.

Each medium frog has a value of 6.

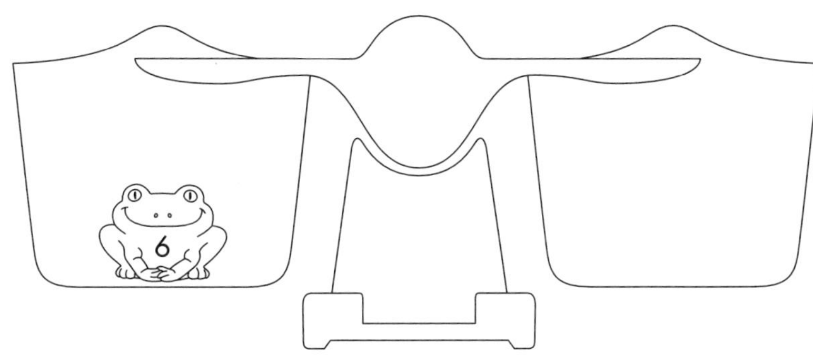

A. What is the value of the frog on the left side of the balance? __________

Put frogs on the right side to balance.
The right side must have the same value as the left side.

B. Write a balanced number sentence.

6 = __________

EXPLORE MORE

Find another way to balance the number sentence.

Find equal values

Name ____________________

Use small and medium frogs.
Record your work.

Each small frog has a value of 3.

Each medium frog has a value of 6.

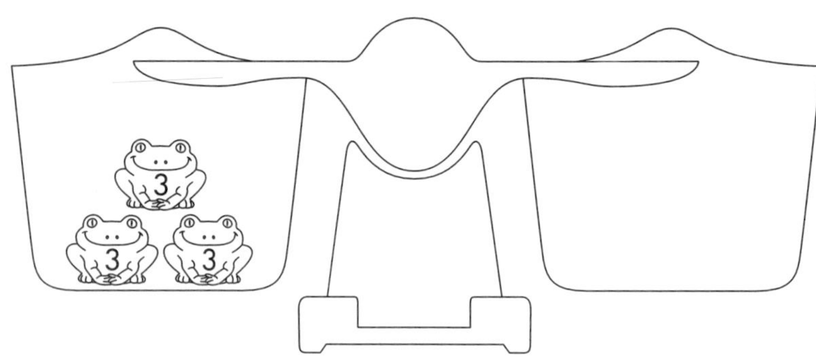

A. What is the value of the frogs on the left side of the balance? __________

Put frogs on the right side to balance.
The right side must have the same value as the left side.

B. Write a balanced number sentence.

3 + 3 + 3 = __________

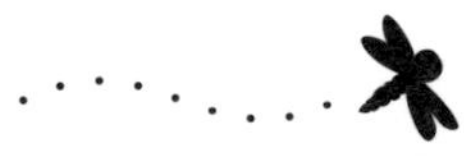

EXPLORE MORE

Find another way to balance the number sentence.

Name ____________________

Use small, medium, and large frogs.
Record your work.

Each small frog has a value of 3.

Each medium frog has a value of 6.

Each large frog has a value of 12.

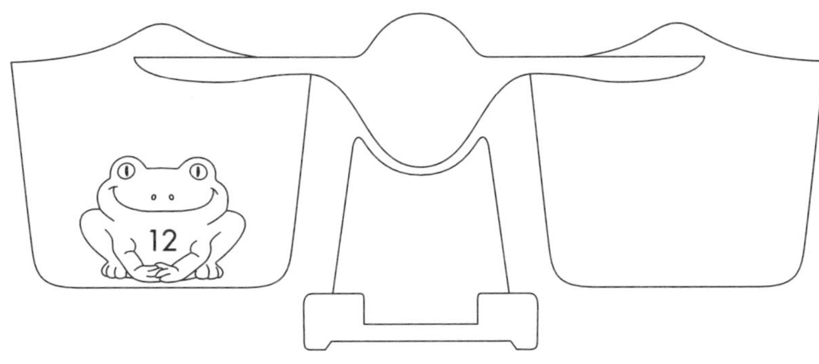

A. What is the value of the frog on the left side of the balance? __________

Put frogs on the right side to balance.
The right side must have the same value as the left side.

B. Write a balanced number sentence.

12 = __________

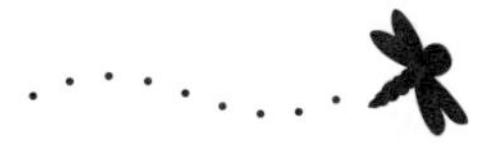

EXPLORE MORE

Find another way to balance the number sentence.

Find equal values

Name ____________________

Use small, medium, and large frogs.
Record your work.

Each small frog has a value of 3.

Each medium frog has a value of 6.

Each large frog has a value of 12.

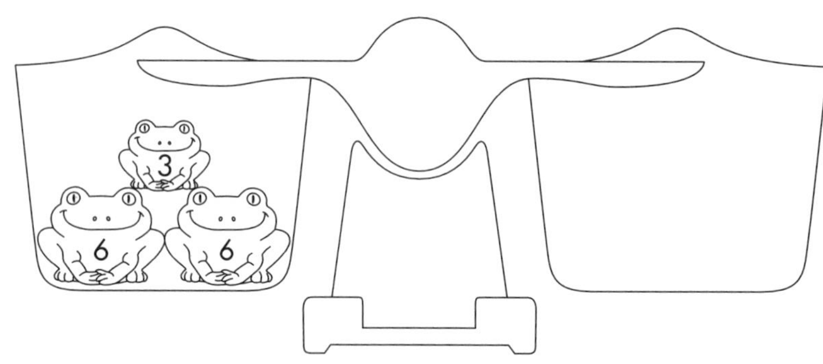

A. What is the value of the frogs on the left side of the balance? __________

Put frogs on the right side to balance.
The right side must have the same value as the left side.

B. Write a balanced number sentence.

6 + 6 + 3 = __________

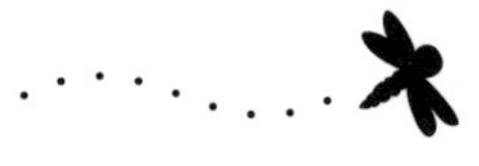

EXPLORE MORE

Find another way to balance the number sentence.

Compare numbers

Name ____________________

Use small, medium, and large frogs.
Record your work.

Each small frog has a value of 2.

Each medium frog has a value of 4.

Each large frog has a value of 8.

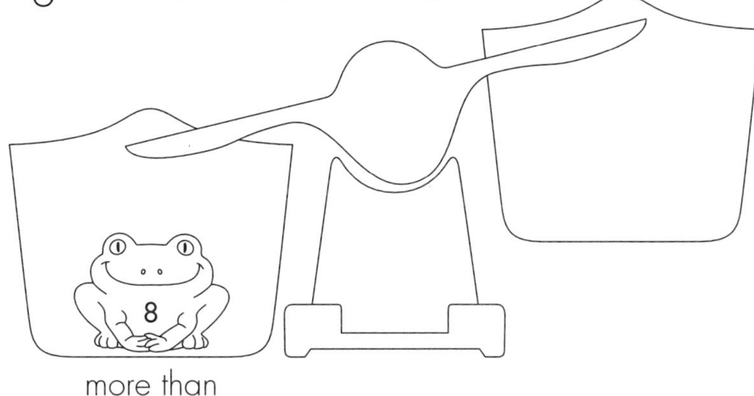

A. What is the value of the frog on the left side of the balance? __________

Put 2 small frogs on the right side of the balance to make the picture true.

B. Fill in the number sentence.

__________ is more than __________

Compare numbers

Name ______________________

Use small, medium, and large frogs.
Record your work.

Each small frog has a value of 2.

Each medium frog has a value of 4.

Each large frog has a value of 8.

more than

A. What is the value of the frogs on the left side of the balance? __________

Put frogs on the right side of the balance to make the picture true.

B. Fill in the number sentence.

__________ is more than __________

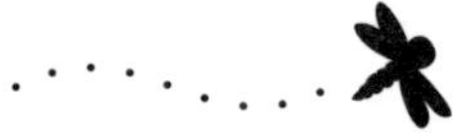

EXPLORE MORE

Find another way to make the picture true.
Fill in the number sentence.

__________ is more than __________

Compare numbers

Name ______________________

Use small, medium, and large frogs.
Record your work.

Each small frog has a value of 2.

Each medium frog has a value of 4.

Each large frog has a value of 8.

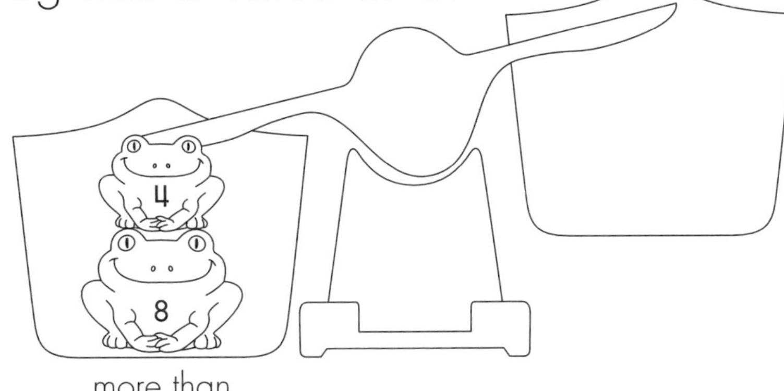

more than

A. What is the value of the frogs on the left side of the balance? __________

Put frogs on the right side of the balance to make the picture true.

B. Fill in the number sentence.

__________ is more than __________

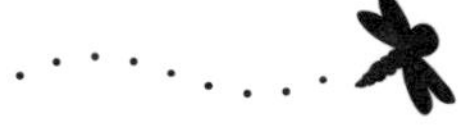

EXPLORE MORE

Find another way to make the picture true.
Fill in the number sentence.

__________ is more than __________

Compare numbers

Name ____________________

Use small, medium, and large frogs.
Record your work.

Each small frog has a value of 2.

Each medium frog has a value of 4.

Each large frog has a value of 8.

more than

A. What is the value of the frogs on the left side of the balance? __________

Put frogs on the right side of the balance to make the picture true.

B. Fill in the number sentence.

__________ is more than __________

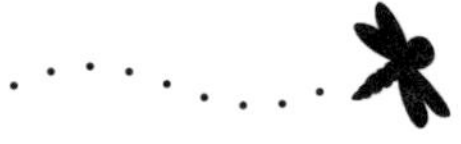

EXPLORE MORE

Find another way to make the picture true.
Fill in the number sentence.

__________ is more than __________

Compare numbers

Name ____________________

Use small, medium, and large frogs.
Record your work.

Each small frog has a value of 2.

Each medium frog has a value of 4.

Each large frog has a value of 8.

A. What is the value of the frogs on the left side of the balance? __________

Put frogs on the right side of the balance to make the picture true.

B. Fill in the number sentence.

__________ is more than __________

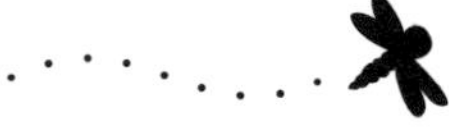

EXPLORE MORE

Find another way to make the picture true.
Fill in the number sentence.

__________ is more than __________

Compare numbers

Name ____________________

Use small, medium, and large frogs.
Record your work.

Each small frog has a value of 2.

Each medium frog has a value of 4.

Each large frog has a value of 8.

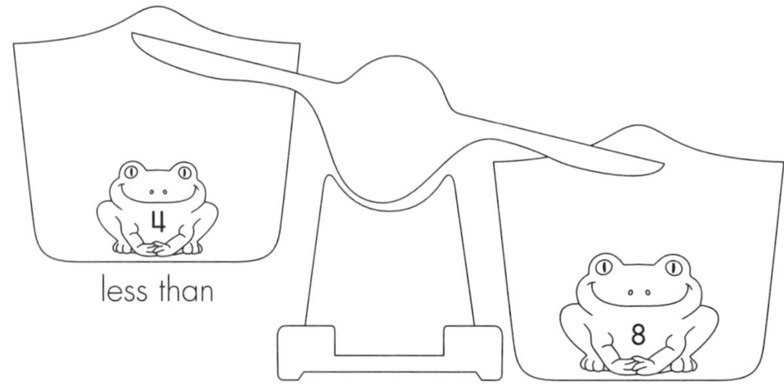

A. What is the value of the frog on the left side of the balance? __________

B. What is the value of the frog on the right side of the balance? __________

C. Fill in the number sentence.

__________ is less than __________

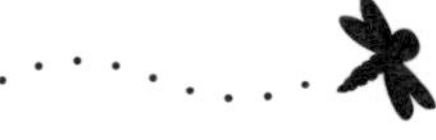

EXPLORE MORE

Find another way to make the picture true.
Fill in the number sentence.

__________ is less than __________

Compare numbers

Name ______________________

Use small, medium, and large frogs.
Record your work.

Each small frog has a value of 2.

Each medium frog has a value of 4.

Each large frog has a value of 8.

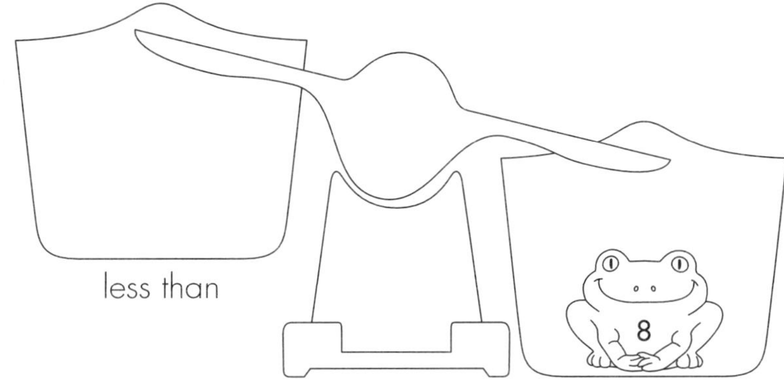

A. What is the value of the frog on the right side of the balance? __________

Put frogs on the left side to make the picture true.

B. What is the value of the frogs on the left side of the balance? __________

C. Fill in the number sentence.

__________ is less than __________

EXPLORE MORE

Find another way to make the picture true.
Fill in the number sentence.

__________ is less than __________

Compare numbers

Name ______________________

Use small, medium, and large frogs.
Record your work.

Each small frog has a value of 2.

Each medium frog has a value of 4.

Each large frog has a value of 8.

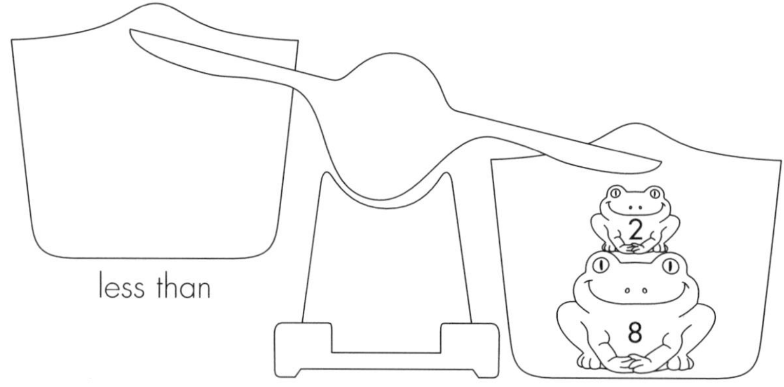

A. What is the value of the frogs on the right side of the balance? __________

Put frogs on the left side to make the picture true.

B. What is the value of the frogs on the left side of the balance? __________

C. Fill in the number sentence.

__________ is less than __________

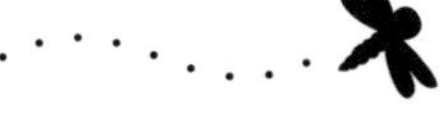

EXPLORE MORE

Find another way to make the picture true.
Fill in the number sentence.

__________ is less than __________

Name ____________________

Use small, medium, and large frogs.
Record your work.

Each small frog has a value of 2.

Each medium frog has a value of 4.

Each large frog has a value of 8.

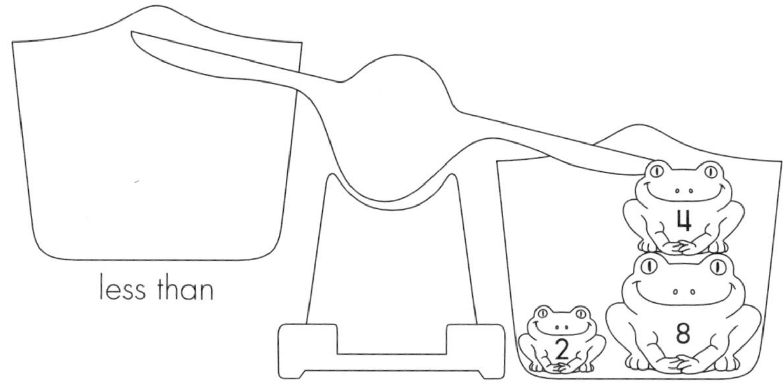

A. What is the value of the frogs on the right side of the balance? __________

Put frogs on the left side to make the picture true.

B. What is the value of the frogs on the left side of the balance? __________

C. Fill in the number sentence.

__________ is less than __________

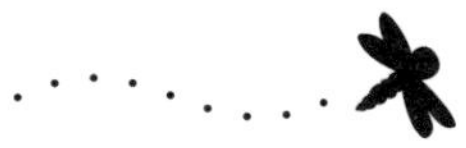

EXPLORE MORE

Find another way to make the picture true.
Fill in the number sentence.

__________ is less than __________

Name ____________________

Use small, medium, and large frogs.
Record your work.

Each small frog has a value of 2.

Each medium frog has a value of 4.

Each large frog has a value of 8.

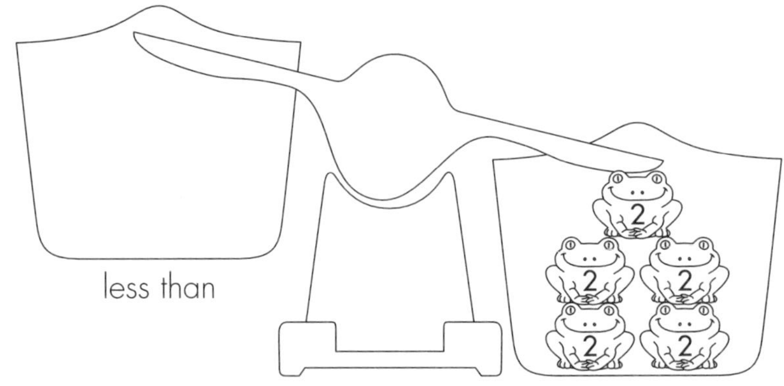

A. What is the value of the frogs on the right side of the balance? __________

Put frogs on the left side to make the picture true.

B. What is the value of the frogs on the left side of the balance? __________

C. Fill in the number sentence.

__________ is less than __________

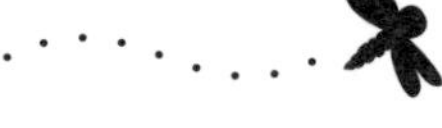

EXPLORE MORE

Find another way to make the picture true.
Fill in the number sentence.

__________ is less than __________

Name ____________________

Use small and medium frogs.
Record your work.

Each small frog has a value of 2.

Each medium frog has a value of 4.

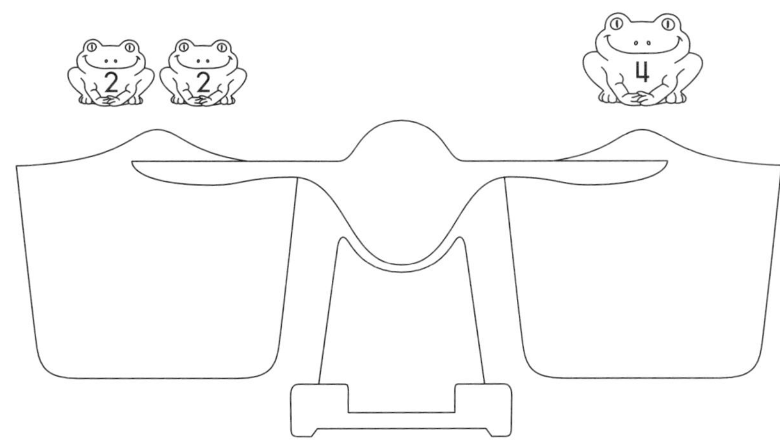

A. Find the value of the frogs on the left side. __________

B. Find the value of the frog on the right side. __________

C. Circle the correct sentence.
Use frogs and a balance to check your answer.

2 + 2 is more than 4

2 + 2 is the same as 4

2 + 2 is less than 4

Compare numbers

Name ______________________

Use small, medium, and large frogs.

Each small frog has a value of 2.

Each medium frog has a value of 4.

Each large frog has a value of 8.

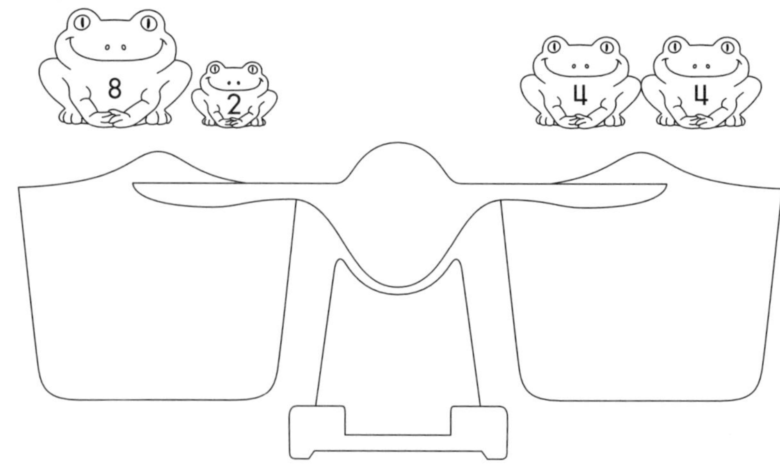

A. What is the value of the frogs on the left side of the balance? __________

B. What is the value of the frogs on the right side of the balance? __________

C. Circle the correct sentence.
Use frogs and a balance to check your answer.

2 + 8 is more than 4 + 4

2 + 8 is the same as 4 + 4

2 + 8 is less than 4 + 4